Larissa C. dos Santos
Danielle Dantas
Natália Baraldi Cunha

Technological development of noodles with nutritious flours:

AF297323

Larissa C. dos Santos
Danielle Dantas
Natália Baraldi Cunha

Technological development of noodles with nutritious flours:

Based on green banana flour and ora-pro-nóbis

ScienciaScripts

Imprint

Any brand names and product names mentioned in this book are subject to trademark, brand or patent protection and are trademarks or registered trademarks of their respective holders. The use of brand names, product names, common names, trade names, product descriptions etc. even without a particular marking in this work is in no way to be construed to mean that such names may be regarded as unrestricted in respect of trademark and brand protection legislation and could thus be used by anyone.

Cover image: www.ingimage.com

This book is a translation from the original published under ISBN 978-613-9-66443-6.

Publisher:
Sciencia Scripts
is a trademark of
Dodo Books Indian Ocean Ltd. and OmniScriptum S.R.L publishing group

120 High Road, East Finchley, London, N2 9ED, United Kingdom
Str. Armeneasca 28/1, office 1, Chisinau MD-2012, Republic of Moldova, Europe
Printed at: see last page
ISBN: 978-620-8-03376-7

Copyright © Larissa C. dos Santos, Natália Baraldi Cunha, Danielle Dantas
Copyright © 2024 Dodo Books Indian Ocean Ltd. and OmniScriptum S.R.L publishing group

SUMMARY

Brazil ranks third among the countries that consume the most pasta, and it is known that the ingredients in traditional pasta do not add important nutritional values to the diet. In view of the fact that Brazilians eat this type of pasta and the possibility of improving it nutritionally, the technical and economic viability of using mixed flours in food has been demonstrated and used in the industry. The aim of this study was to make noodles using mixed green banana and ora-pro-nobis flours and to characterise them nutritionally, as well as applying the **sous vide** technique in order to improve their nutritional quality and increase their usual shelf life. To this end, green banana and ora-pro-nobis flour was produced. The proportions of the ingredients used were: 220g of green banana flour, 50g of ora-pro-nobis flour, 150g of white flour, 4g of guar gum, 4g of xanthan gum, 1 tablespoon of olive oil, 4 eggs, 8g of salt and 8 tablespoons of water, which were mixed together to make the dough, which was passed through a cylinder at 3mm for a noodle-like cut. After making the pasta, 6 samples were separated and subdivided into 2 groups, with sauce and without sauce. These were pre-cooked for around 40 seconds and vacuum-packed, then cooked in a sous **vide** thermocirculator for 4 minutes at a temperature of 73°C, after which they were frozen at <12°C for 5, 15 and 30 days consecutively. To analyse macronutrients, the moisture content of the sample was quantified, which was taken to an oven at 105°C and then calculated using the formula **u=mi-mfx100/mi, for the** ash **content**, the sample was incinerated in a muffle furnace at 550°C for 2 hours; for the protein content, this was carried out using the Kjeldahl method, which is based on transforming the nitrogen in the sample into ammonium sulphate through digestion with sulphuric acid and subsequent distillation with the release of ammonia, which was fixed in an acid solution and titrated, the calculation used being %N=0.014xVxfxMx100; for the lipid analysis, the samples were weighed (around 5 g) in a Soxhlet cartridge and then transferred to a Soxhlet extractor and then 150 ml of petroleum ether was added to the flat-bottomed flasks and heated at between 35°C and 40°C for 6 hours, after which the flasks were transferred to an oven at 105°C for 1 hour, the calculation used to obtain the value was L=100xN/P.

The carbohydrate content was calculated using the formula %Carbohydrate=100-% Moisture-% Ash-%Lipids-%Protein-%Fibre. The results obtained were positive for the samples soaked for 5 and 15 days, but for the samples without soaking, regardless of the period, and 30 days, with and without soaking, they were partially positive, as they were dry and brittle. The **sous vide** technique seems to have preserved the organoleptic characteristics of the product, which had a more intense colour compared to traditional cooking.

Keywords: Green banana flour. Ora-pro-nóbis flour. PANC. Pasta. **Sous Vide.**

CHAPTER 1

INTRODUCTION

Over the last 20 years, Brazil and Latin American countries have gone through a period of nutritional transition marked by an increase in the intake of fatty, refined and sugar-rich foods and a low consumption of fruit and vegetables (BATISTA FILHO & RISSIN, 2003). With the increased consumption of processed and ultra-processed foods and the low consumption of fresh foods, the majority of the population does not meet the recommended daily intake of fibre, vitamins and minerals, as shown in the latest Family Budget Survey (Pesquisa de Orçamento Familiar, PO, 2008-09) published by the Brazilian Institute of Geography and Statistics (Instituto Brasileiro de Geografia e Estatística, 2010).

Among the most varied types of food consumed by the population, Brazil ranks third among the countries that consume the most pasta, with 1.2 million tonnes, second only to Italy and the United States (ABIMAP, 2015). According to the National Health Surveillance Agency (ANVISA) resolution RDC number 14 of 21 February 2000 (2000), pasta is defined as a non-fermented product obtained by kneading wheat flour, semolina or wheat semolina with water, with or without the addition of other permitted substances. ANVISA also classifies noodles according to their moisture content as dry or fresh, depending on their moisture content.

The ingredients in conventional noodles do not add important nutritional values to the diet, as they contain 77.9 grams (g) of high glycaemic index carbohydrates, which contributes to an increase in blood glucose levels, low amounts of fibre, vitamins and minerals, offering 371 calories (Kcal) per 100 grams of pasta (TACO, 2011).

Given the popularity of pasta among Brazilians, its ease of preparation and the possibility of improving it nutritionally, the technical and economic viability of using mixed flours in food is being demonstrated and used in the industry. Mixed flours used in pasta preparation can be of the starchy type, which contains high levels of starch, such as corn flour, cassava, yam, green banana, etc., and proteinaceous, when they contain high levels of protein, such as flours from soya, legumes and others (EL-

DASH; GERMANI, 1994).

While still green, bananas have a significant content of resistant starch (RA) (RAMOS; LEONEL; LEONEL, 2009), which has been shown to be beneficial in lowering the glycaemic index, reducing the risk of cardiovascular diseases, as well as being related to weight loss (RAIMUNDO et al., 2013). The drying of green bananas and their subsequent grinding produces a flour from the fruit, characterising a starchy type flour, which has better nutritional value compared to other flours and can be used as a raw material for various products and/or as a substitute for other flours, especially in pasta (EMBRAPA, 2009).

The ora-pro-nóbis, **known as "lobrobo", "meat of the poor" and "vegetable meat", has a high protein content** (ARDISSONE, 2009) and is considered easy to grow and propagate. It can be used both raw and processed in various preparations such as salads, stir-fries, pies and pasta such as macaroni, after dehydration and grinding to produce a proteinaceous flour (ALMEIDA, 2014).

Nowadays, food companies have become even more concerned about preserving their food in order to minimise the damage caused by easy deterioration, especially when it comes to homemade pasta (RAMOS; LEONEL; LEONEL, 2009). **Sous vide** is a technique that has been used in which food is previously vacuum-packed and then heat-treated in plastic bags (RAMOS, 2004). This technique allows the food to maintain its aspects such as: nutritional value, colour, flavour and appearance, as well as providing greater durability to the preparation, as the vacuum prevents the food from being degraded by aerobic microorganisms (MEDEIROS, 2009).

With this in mind, the aim of this study was to make noodles using mixed green banana and ora-pro-nobis flours, to characterise them nutritionally in terms of macronutrient content and organoleptic characteristics, appearance and colour, and to apply the **sous vide** technique to determine the durability of the food produced, in order to improve the nutritional quality of the noodles and increase their usual shelf life.

1.1GREEN BANANA

Bananas, the fruit of the banana tree, are one of the most widely grown fruits in

Brazil, with a production of 6,892,622 million tonnes in an area of 486,991.00 thousand hectares, making it the fourth largest producer in the world (FOOD AND AGRICULTURE ORGANISATION, 2013). On the national scene, Bahia's 1,113,93 tonnes and São Paulo's 1,090,009 tonnes stand out (IBGE, 2013).

Considered one of the most important fruits in the world, both in terms of production and marketing, bananas are not only a complementary food in the diet of the Brazilian population, but are also widely accepted due to their sensory characteristics and their ease of social and economic consumption, with consumption of 11.4kg/inhabitant/year, second only to oranges with 12.2kg/inhabitant/year (FOOD AND AGRICULTURE ORGANISATION, 2013).

In addition, it is used as a source of income for many families by large, medium and small producers, generating job opportunities in the countryside and in the city, contributing to the development of the regions involved in its production (SILVA et al., 2004).

Climatic conditions such as temperature, relative humidity, rainfall and sunshine favour the fruit's year-round production, which is typical of tropical countries. Brazil is a major world producer of the fruit, but there is a high rate of wastage, with losses of up to 60 per cent in the pre- and post-harvest processes (LICHTEMBERG, 2008).

At this stage of development, the fruit has a high water content and a high amount of starch. Only when ripe does the fruit acquire a yellowish colour, brown spots, a soft consistency and a sweeter taste (BRAZILIAN SERVICE TO SUPPORT MICRO AND SMALL ENTERPRISES, 2008).

This fruit has important nutritional content such as vitamins A, C, B1, B2, mineral salts such as potassium, phosphorus, sodium, magnesium and calcium. It also contains three types of natural sugar, sucrose, fructose and glucose, which when combined provide the body with great energy (RAIMUNDO et al., 2013).

While the fruit is still green, it has a significant content of resistant starch (RA), whose values vary on average from 10 to 40 per cent, depending on the banana genotype (RAMOS; LEONEL; LEONEL, 2009), which has been shown to be beneficial in lowering the glycaemic index, reducing the risk of cardiovascular diseases, as well as being related to weight loss (RAIMUNDO et al., 2013).

Green bananas can be used to make flour, which is obtained by naturally or artificially drying the pulp of the green or semi-green fruit and then grinding it. It has a better nutritional value than other flours and can be used as a raw material for various products, especially pasta and/or to replace other flours in a wide variety of preparations (EMBRAPA, 2009).

1.2. BLACKCURRANT

Brazil has the largest tropical forest cover in the world, especially concentrated in the Amazon Region. Combined with its territorial extension, geographical and climatic diversity, our country is home to an immense biological diversity, which makes it the leading megadiverse country on the planet, with between 15% and 20% of the 1.5 million species described on Earth. It has the richest flora in the world, with around 55,000 species of higher plants (approximately 22 per cent of the world total); 56,000 species of higher plants, including 162 species of cacti, of which 123 are endemic (LEWINSOHN E PRADO, 2000).

Amongst this diversity of species is ora-pro-nobis, a vegetable considered unconventional by the Brazilian government (BRASILÍA, 2010). This type of vegetable, which also goes by the name of **unconventional food plant** (PANC), has become an alternative for taking advantage of Brazil's biodiversity, rescuing and valorising it, representing important gains from a cultural, economic, social and nutritional point of view, considering the tradition of cultivation by various communities (BRASILÍA, 2010). The term **"unconventional"** is used **to refer to those** species that have not yet received due attention from the technical-scientific community and society as a whole, resulting in localised consumption in certain localities or regions, with difficulty in penetrating the rest of the country. (BRASILÍA, 2010).

According to Knupp and Barros (2008), unconventional fruit and vegetables generally have significantly higher mineral and protein contents than domesticated plants, as well as being richer in fibre and compounds with functional roles.

The ora-pro-nobis (Pereskiaaculeta), from the Latin **meaning "Pray for us"**, belongs to the cactaceae family and is one of the only plants with lance-shaped, succulent leaves. It is a climbing plant with white flowers and small, yellow fruits that can be used to make jellies, liqueurs and ice creams. They originated in America, but are widely cultivated in Brazil and are present in the cuisine of Minas Gerais in the municipality of Sabará, forming part of the eating habits of the population of this region (BRASILÍA, 2010). Known as **"lobrobo", "Meat of the Poor" and "Vegetable Meat", it has a** high protein content (25% of its total calories), 85% of which is in digestible form, easily used by the body, and contains some essential amino acids, including lysine, which is essential for strengthening the immune system. It is also rich in vitamins A, B, C, fibre, iron and phosphorus and is a source of mucilage (ARDISSONE, 2009).

The ora-pro-nobis is considered easy to grow and its propagation has a low water demand and low incidence of disease, favouring domestic cultivation. Its leaves can be used both raw and processed, being used in various preparations such as flours, salads, stir-fries, pies and pasta such as macaroni (ROCHA et al., 2008).

1.3 CONTEMPORARY *SOUS VIDE* METHOD

In order to preserve the nutritional properties of the product developed, the **sous vide** technique can be used as a packaging method. According to Medeiros (2009), this type of technique allows the food to maintain its aspects such as: nutritional value, colour, flavour and appearance and provides greater durability, as the vacuum prevents the food from being degraded by aerobic microorganisms.

Sous vide is a technique in which food is previously packaged in plastic vacuum bags, in which it can be inserted raw or pre-cooked, and then undergoes heat treatment in these same bags. When temperatures reach between 60° and 90°C, the product is taken to a cold room to be cooled and stored at -25°C, and can be used for up to 18 months (RAMOS, 2004).

The **sous vide** technique is mainly used for cooking food at a low temperature for a long period of time. This procedure maintains flavours, improves texture and preserves the food's natural humidity (GARCIA E HEBBEL, 2010). The use of this technique is an important method for determining the durability of noodles enriched with a mix of green banana flour and ora- pro-nobis, as the food developed does not contain any chemical preservative additives and the technique allows air to be removed from the packaging, as well as cooking for a prolonged period of time, preventing the proliferation of microorganisms.

CHAPTER 2

METHODOLOGY

The methods used to make the pasta with the flour mix are described below:

2.1. PRODUCTION OF GREEN BANANA FLOUR

The raw material used was green bananas of the **Cavendish** group, the nanica variety, in the green stage, with grade 1 ripeness. The fruit was selected and washed under running water, weighed on an SF-400 digital scale, cut and submerged in citric acid solution, dried in an oven at 80°C and then processed in a Skymsen industrial blender. The procedures were carried out in the Nutrition Laboratory at the University of the Sacred Heart (USC), based on the methodology described by Souza (2009), with the necessary alterations, which will be mentioned in the course of the research. Figure 1 shows the flowchart used to obtain green banana flour.

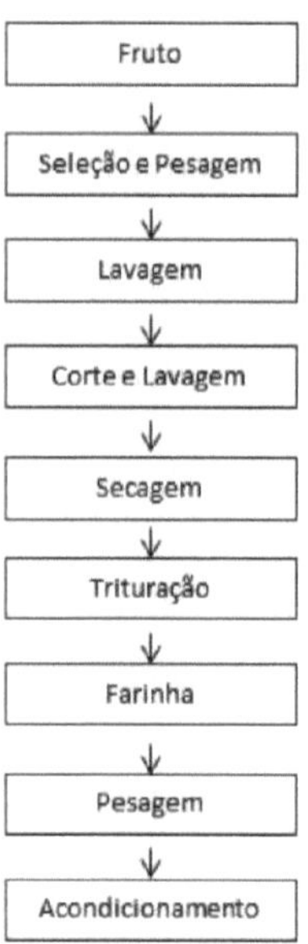

Flowchart 1 - Flowchart for obtaining green banana flour.

2.1.1 Sorting and Weighing

The samples acquired were selected according to size and degree of ripeness, then weighed (APPENDIX A) for subsequent analysis of the flour yield.

2.1.2 Washing

To remove dirt from the surface of the banana peel, the selected fruits were washed one by one in running water. It was not necessary to chlorinate the fruit due to the high temperature it was subjected to in order to obtain the dried fruit.

2.1.3 Cutting and sulphiting

The bananas, with the peel on, were sliced into slices approximately 1 mm to 2 mm thick. They were then subjected to antioxidant treatment with citric acid to prevent the slices from darkening

(APPENDIX B). In this process, the slices were immersed in a solution of citric acid at a concentration of 6 grams (g) to 1 litre (L) of water. These procedures were based on the methodology of the study by Souza (2009), but with a higher concentration because it is a natural acid. The slices were kept immersed in the solution for 19 hours under refrigeration.

2.1.1. Drying

After the slices had been kept in the citric acid solution for 19 hours, they were removed from the solution and left on absorbent sheets to remove excess water, after which they were placed on screens suitable for carrying out the dehydration process (APPENDIX C). They were then transferred to an oven with air circulation at a constant temperature of 80°C for four hours.

2.1.2. Crushing, Weighing and Packaging

The dehydrated slices were ground in an industrial blender and sieved through a steel mesh, then weighed to analyse the yield. After obtaining the flour, it was packed in plastic polythene bags and stored at room temperature until it was used to make pasta.

2.2. PRODUCTION OF ORA-PRO-NÓBIS FLOUR

The raw material used was the PANC from the cactaceae family, ora-pro- nobis (pereskiaaculeta), which was washed in running water, weighed on a SF-400 scale and processed in a Skymsen industrial blender. These procedures were carried out at the USC Nutrition Laboratory, based on the methodology described by Rocha et al. (2008) for obtaining flour, with the appropriate changes, which will be mentioned in the course of the research. Figure 2 shows the flowchart used to obtain ora-pro-nóbis flour.

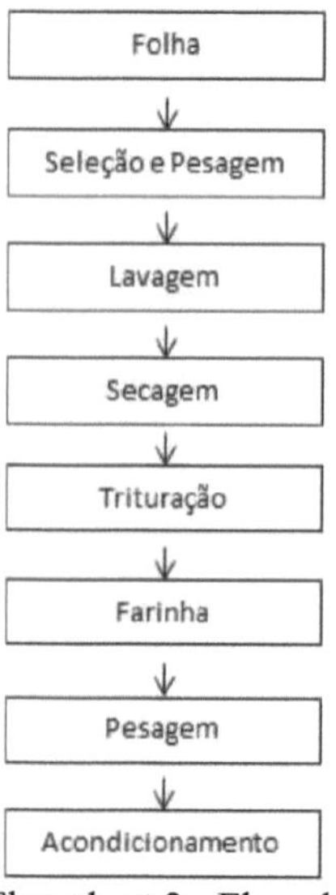

Flowchart 2 - Flowchart for obtaining ora-pro-nobis flour.

2.2.1. Sorting and weighing

The samples acquired were taken from randomly chosen plants and then selected according to their physical preservation, after which they were weighed for subsequent yield analysis (APPENDIX D).

2.2.2. Sanitising and washing

To remove dirt from the surface of the leaves, they were sorted and washed one by one in running water. Chlorination was not necessary due to the high temperature they were subjected to in order to obtain the flour.

2.2.3 Drying

After sanitising and washing, the leaves were placed in screens suitable for the dehydration process (APPENDIX E). They were then transferred to an oven with air circulation at a constant temperature of 65°C until they were completely dehydrated, with an estimated time of 2 hours (APPENDIX F).

2.2.4 Crushing, weighing and packaging

The dehydrated leaves (APPENDIX G) were ground in an industrial blender and sieved through a steel mesh, then weighed to analyse the yield. Once the flour had been obtained, it was packed in plastic polythene bags and stored at room temperature until it was used to make noodles.

2.3   MAKING THE NOODLES

2.3.1 Ingredients   used to make pasta dough

The following ingredients were used to make the noodles: wheat flour, green banana flour, ora-pro-nobis flour, guar gum, xanthan gum, eggs, salt, olive oil and water. Figure 3 shows the ingredients and quantities used to make the noodles.

Table 1 - Green banana and ora-pro-nobis flour noodle recipe

Green Banana Flour and Ora-pro-nobis Noodles

Ingredients	Quantity
White Wheat Flour	150g
Green Banana Flour	220g
Ora-pro-nobis flour	50g
Gomar Guar	4g
Xanthan gum	4g
Eggs	4 und.
Olive oil	1 tablespoon
Water	8 tablespoons
Salt	8g

2.3.2 Making the noodles

The pasta was produced based on a traditional noodle recipe, in which 400g of wheat flour is used. However, in this new recipe, part of the wheat flour was replaced with green banana and ora-pro-nobis flour. In addition, 1% of each gum (guar and xanthan) was added in order to structure and improve the quality of the pasta in relation to the amount of green banana and ora-pro-nobis flour used, according to Pires (2012).

To make the noodles, the ingredients were separated and weighed in the appropriate proportions, then the solids such as green banana flour, ora-pro-nobis flour, white flour, salt, guar gum and xanthan gum were mixed in a Bolw.Then the eggs, olive oil and water were added. After the desired consistency had been reached, the product was wrapped in plastic food wrap and the dough was left to rest for 20 minutes at room temperature.

After resting, the pasta was cut in an industrial cylinder to a thickness of 1 mm in a noodle mould. Next, wheat flour was sprinkled over a stainless steel vat, in which the pasta was left to dry for 10 minutes. The pasta was then boiled in water with olive oil at a ratio of 100g of pasta to each litre of water.

2.2.3. Making tomato sauce

To enrich the product nutritionally and visually, a tomato sauce was made using the following ingredients: tomato, onion, garlic, Calabrian chilli, salt and olive oil. Figure 4 shows the ingredients and quantities used to make the sauce.

Table 2 - Tomato sauce recipe

Tomato sauce	
Ingredients	**Quantity**
Tomatoes	+ 10 medium units (1 kg)
Onions	1 medium unit (0.160g)
Garlic	3 cloves of garlic
Salt	1 teaspoon
Calabrian chilli	1 pinch (0.001 g)
Olive oil	1 tablespoon

To prepare the sauce, the tomatoes were first sanitised under running water, without the need to chlorinate them, as they had been subjected to cooking. After cleaning, the concassé process was applied, which consists of removing the stalk from the tomato and making a shallow X-shaped cut on the side opposite it, then immersing the tomatoes in boiling water for approximately 3 minutes until their skin is detached from the flesh of the fruit and then applying heat shock in order to stop them cooking and cool them down so that the skin can be removed more easily and the seeds removed. After the concassé process, the tomatoes were cut into cubes and cooked. To do this, the olive oil, onion and garlic were added to a hot pan to sweat and then the tomatoes were added along with the salt, leaving the mixture to simmer for approximately 30 minutes. Using a mixer, the tomatoes were blended and then the Calabrian chilli was added to the sauce.

2.4.   SOUS VIDE METHOD

According to Rocha et al. (2008) and Baldwin ([2000]), the sous *vide* method has three stages: preparation for packaging, cooking and finishing. The equipment needed for this is a vacuum packer and a thermocirculator, which were provided by the USC Gastronomy laboratory.

2.4.   1Pre-coction

The pasta was pre-cooked for 40 seconds to prevent the noodle strands from sticking together and then subjected to thermal shock (running water).

2.4.2Preparing   the packaging

The noodles were divided into 6 portions of 100g and subdivided into 2 groups, samples with sauce and without sauce. After separation, they were individually packed in plastic Nylon Poly bags, whose temperature range is -30°C to 100°C, identified and then subjected to vacuum packing to remove the air.

2.4.   3Cooking

Cooking was carried out using the sous vide thermocirculator for 4 minutes at around 73°C and then rapidly cooled for freezing.

2.4.   4Finalisation

The samples were kept frozen at <12°C for 5, 15 and 30 days consecutively to determine the appropriate shelf life and changes in organoleptic characteristics.

2.5.   TECHNOLOGICAL ANALYSIS

The methodology used to determine the nutritional value of the product is described below:

2.5.1Sampling   and sample maintenance

A simple random sample of 100g of the dough was taken to represent the product. The sample was ground in a blender and

reserved in a jar labelled with the type of material and stored at room temperature.

2.5.   2Humidity

To determine the moisture content of the product, the sample was dried in an oven at 105 °C. The porcelain crucible was placed in the oven for 1 hour to remove all moisture. After drying, the crucibles were weighed on an analytical balance. Moisture was determined in an oven at 100 °C overnight. After this period, the samples were placed in a desiccator and, once cooled, weighed again on an analytical balance (INSTITUTO ADOLFO LUTZ, 2008).

The moisture content is calculated using the following formula:

$$U = \frac{mi-mf}{mi} \times 100$$

U = humidity (%)

mi= initial sample mass (g) mf= final sample mass (g).

2.5.3 Ash

The materials needed for this analysis were: porcelain crucibles, analytical balance, desiccator, muffle furnace (550 °C), spatula and tweezers. This method consists of the total incineration of the sample without smoke being released, followed by heating in a muffle furnace. It is carried out in such a way as to completely incinerate the sample without any smoke being released, followed by heating in a muffle furnace at 550°C for 2 hours. The sample was completely incinerated and a charcoal-free residue with a greyish colour was obtained (HEIDEIN et al., 2014). All the analyses were carried out in triplicate, with the results expressed as a function of the average ash content (in percentage), standard deviation and coefficient of variation.

The ash content is calculated using the following formula:

$$C\ (g) = pf - pi$$

pf = final crucible weight

pi= initial weight of crucible (empty)

$$\% \ cinza = c\ (g) \times 100/\ pa$$

pa = sample weight.

2.5. 4Proteins

The determination of protein in a sample is based on the determination of nitrogen. The Kjeldahl method is based on transforming the nitrogen in the sample into ammonium sulphate through digestion with sulphuric acid and subsequent distillation with the release of ammonia, which is fixed in an acid solution and titrated. All the analyses were carried out in triplicate, with the results expressed as a function of the average protein content (in percent), standard deviation and coefficient of variation. 0.2 g of the sample, 10 glass beads, 5 mL of sulphuric acid (H_2SO_4) and 2.5 g of catalyst mixture (96% potassium sulphate + 4% copper sulphate) were added to the Kjedahl tubes. These were placed in the digester for 4 hours, gradually increasing the temperature until it reached 400°C. After digestion, the tubes were placed in the nitrogen distiller. Then 7 mL of sulphuric acid and 20 mL of sodium hydroxide (NaOH) were added to the tubes. 10 mL of boric acid was pipetted into the flask and 2 drops of the mixed indicator (bromocresol green and methyl red in alcoholic solution) were added, and finally the resistor was switched on to bring the water to the boil. In the chemical reaction in the Kjedahl tube, the salt ammonium sulphate was formed. The sodium hydroxide fell into the Kjedahl tube, breaking the covalent bonds in the salt and releasing the ammonia, which was dragged with the water through the distiller until they condensed. The ammonia fell into the flask containing boric acid (H_3BO_3), forming ammonium borate which, in the presence of the mixed indicator, changes colour from pink to green.

After distillation, the ammonium borate was titrated with hydrochloric acid (HCl) at a concentration of 0.1M; the colour changed from green to light pink.

The ash content is calculated using the following formula:

$$\% N = 0{,}014 \times V \times f \times M \times 100$$

V = volume of 0.1 N sulphuric acid solution or 0.1 N hydrochloric acid solution used in the titration after correcting the blank, in mL;

N = theoretical normality of the 0.1 N sulphuric acid solution or 0.1 N hydrochloric acid solution;

f = correction factor of 0.1 N sulphuric acid solution or 0.1 N hydrochloric acid solution;

m = sample mass, in grams.

The Kjeldahl method quantifies the total nitrogen in the sample. To obtain the protein content, a factor called the **"Kjeldahl conversion factor"** is applied **(INSTITUTO ADOLFO LUTZ, 2008).**

$$\% \ protein = \% \ total \ nitrogen \ \text{x} \ F$$

F = nitrogen/protein ratio conversion factor, F = 6.38.

As the determination is made in duplicate, express the result of the average (% protein content) to one decimal place.

2.5.5. Lipids

All analyses were carried out in triplicate. The samples were weighed (about 5 g) into a Soxhlet cartridge. The cartridges were transferred to the Soxhlet extractor and then the flat-bottomed flasks were pre-weighed at 105°C for 1 hour and cooled in the desiccator. After the flasks had cooled, they were weighed empty. 150 ml of petroleum ether was then added to the flasks and then heated to a temperature of between 0.5 and 0.5°C.

At 35°C and 40°C for 6 hours until the solvent evaporated, the flasks were then transferred to an oven at 105°C for 1 hour and then cooled in a desiccator. Finally, the flasks were weighed after the procedures to obtain the weight of the fat retained in the container. The calculation of the lipid content is obtained using the following formula:

$$L = \frac{100 \times N}{P}$$

N = number of grams of lipids

P = number of grams of sample

2.5.   6Carbohydrates

Carbohydrates were determined after the other nutrients had been determined using the formula:

% Carbohydrate = 100 - % Moisture - % Ash - %Lipids - %Protein - % Fibre

CHAPTER 3

RESULTS

The results obtained are described below:

3.1. GREEN BANANA FLOUR

After selecting, washing, cutting, immersing in an antioxidant solution, then drying and crushing the fruit, a light-coloured flour-like product was obtained, with no apparent odour or taste. The appearance of the flour can be seen in figure 1.

Figure 1. Green banana flour.

3.2. ORA-PRO-NOBIS FLOUR

After sorting, washing, drying and grinding the leaves, a product was obtained with a mossy green texture and a characteristic taste and smell. The appearance of the flour can be seen in figure 2.

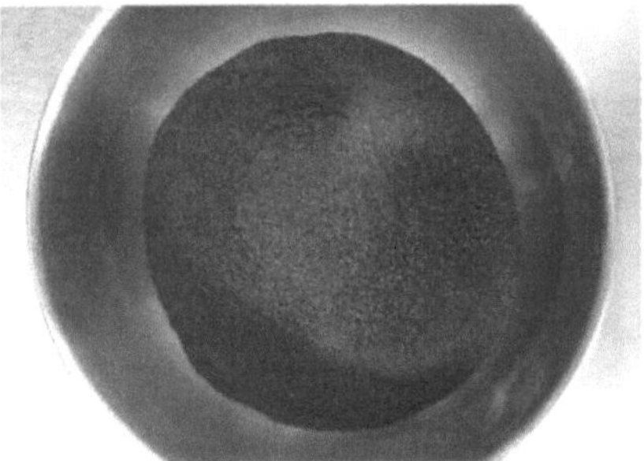

Figure 2: Prunus flour

3.3 NOODLES

After combining the ingredients mentioned in figure 2, a dough was obtained that was suitable for handling to form noodles. It was firm, had satisfactory elasticity and was not brittle after the addition of the gums (guar and xanthan) and the ora-pro-nobis flour adhered well to the dough, as can be seen in figure 3.

Figure 3: Noodles made with green banana flour and ora-pró-nobis.

3.4. MACRONUTRIENT ANALYSIS

3.4.1 Moisture analysis

Table 3 - Moisture analysis data.

filter weight n°	Empty filter weigher	Initial sample	Final weight	Final sample
1	41,581g	3,498g	44,789g	3,208g
2	38,754g	3,494g	41,959g	3,206g
3	41,473g	3,509g	44,691g	3,218g

Source: prepared by the authors.

Calculations:

-Filter weight 1:

$$U= \frac{3,498g - 3,208g}{3,498g} \times 100$$

$$= 8,29\%$$

-Filter weight 2:

$$U = \frac{3,494g - 3,206g}{3,494g} \times 100$$

$$= 8,24\%$$

- Filter weight 3:

$$U = \frac{3,509g - 3,218g}{3,509g} \times 100$$

$$= 8,29\%$$

From the results obtained above, it was possible to find the following values in Excel: Mean, Standard Deviation and Coefficient of Variation. All the values found were used because they were very close.

Table 1 - Mean, standard deviation and coefficient of variation for humidity.

Average	Standard Deviation	Coefficient of Variation
8,27%	0,3	0,35

The average moisture content found in the sample was 8.27 per cent.

3.4.2 Grey

Table 4 - Data for the ash analysis.

crucible no.	Empty crucible	Initial sample	Final crucible weight	Final sample
1	27,230g	3,510g	27,405g	0,175g
2	27,349g	3,519g	27,531g	0,182g
3	30,230g	3,513g	30,411g	0,181g

Calculations:

- Crucible 1:

$$\%C = 0.175 \times 100 / 3.510 = 4.98\%$$

- Crucible 2:

$$\%C = 0.182g \times 100 / 3.519g = 5.17\%$$

- Crucible 3:

$$\%C = 0.181g \times 100/3.513 = 5.15\%$$

From the results obtained above, it was possible to find the following values in Excel: Mean, Standard Deviation and Coefficient of Variation. All the values found were used because they were very close.

Table 2 - Mean, standard deviation and coefficient of variation for ash.

Average	Standard Deviation	Coefficient of Variation
5,10%	0,10	2,05

The average ash content found in the sample was 5.10 per cent.

3.4.3 Protein

Table 5 - Quantity, in grams (g), of samples for protein analysis.

No. Tube	Sample
1	0,199g
2	0,206g
3	0,201g

Table 6 - Values used for titration.

No. Tube	Titrated volume
1	3.2ml
2	3.1ml
3	3.0ml

Calculations:

- Tube 1:

$$\%N = \frac{0.014 \times 3.2 \times 0.93 \times 0.1 \times 100}{0,199}$$

$$\%N = 2,09\%$$

$\%P = 2,09 \times 6,25$

$\%P = 13,06\%$

- Tube 2:

$$\%N = \frac{0{,}014 \times 3{,}1 \times 0{,}93 \times 0{,}1 \times 100}{0{,}206}$$

$$\%N = 1{,}95\%$$

%P= 1,95 x 6,25

%P= 12,18%

- Tube 3:

$$\%N = \frac{0{,}014 \times 3{,}0 \times 0{,}93 \times 0{,}1 \times 100}{0{,}201}$$

$$\%N = 1{,}94$$

%P = 1,94 x 6,25

%P = 12,12%

From the results obtained, it was possible to find the following values in Excel: Mean, Standard Deviation and Coefficient of Variation. All the values found were used because they were very close.

Table 3 - Mean, standard deviation and coefficient of variation for protein.

Average	Standard Deviation	Coefficient of Variation
12,45%	0,53	4,23

Source: prepared by the author.

The average protein content found in the sample was 12.45 per cent.

3.4.4 Lipids

Table 7 - Data for the lipid analysis.

n° Balloon	Empty balloon	Initial sample	Final weight of Balloon	Final sample
1	102,342g	5,831g	102,750g	0,408g
2	113,637g	4,944g	113,970g	0,333g
3	102,953g	4,974g	103,300g	0,347g

Source: prepared by the authors.

Calculations:

B Balloon 1:

$$\%L = 0.408g \times 100/5.381 = 7.58\%$$

- Balloon 2:

$$\%L = 0.333g \times 100/4.944 = 6.73\%$$

- Balloon 3:

$$\%L = 0.347g \times 100/4.974g = 6.97\%$$

From the results obtained, it was possible to find the following values in Excel: Mean, Standard Deviation and Coefficient of Variation. However, only the values for flasks 1 and 3 were used, as flask 2 was discarded due to the fact that the ether had evaporated, not refluxing and thus burning the sample, as shown in the figure below (Figure 4).

Table 4 - Lipid mean, standard deviation and coefficient of variation.

Average	Standard Deviation	Coefficient of Variation
7,28%	0,43	5,93

Figure 4 - Bottom of flask 2 with burnt lipid sample.

The average lipid content found in the sample was 7.28 per cent.

3.4.5 Carbohydrate

$$\%CHO = 100 - \%8{,}27 - \%5{,}10 - \%7{,}28 - \%12{,}45$$
$$\%CHO = 66.9\%$$

The percentage of fibre was not used to calculate carbohydrates as it could not be quantified. However, it is worth noting that fibre is part of a group of carbohydrates. The average carbohydrate content found in the sample was 66.9%.

1.5 HOMEMADE TOMATO SAUCE

The results were satisfactory, as it had a characteristic colour and appearance, and there was no need to add food colouring, even natural food colouring, to enhance the colour (Figure 5).

Figure 5: Homemade tomato sauce.

3.6 VACUUM PACKAGING

The results obtained were partially positive for the samples packaged without the presence of sauce, since when they were opened after the packaging process, the strands stuck together and were brittle, while the samples packaged with the presence of tomato sauce showed positive results, since when they were opened, the strands were loose and less brittle compared to the samples without sauce (Figures 6 and 7).

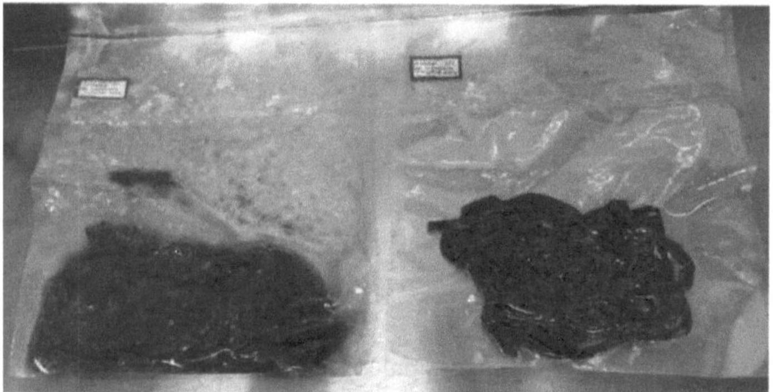

Figure 6. Vacuum-packed samples of green banana and ora-pro-nobis flour noodles with and without sauce.

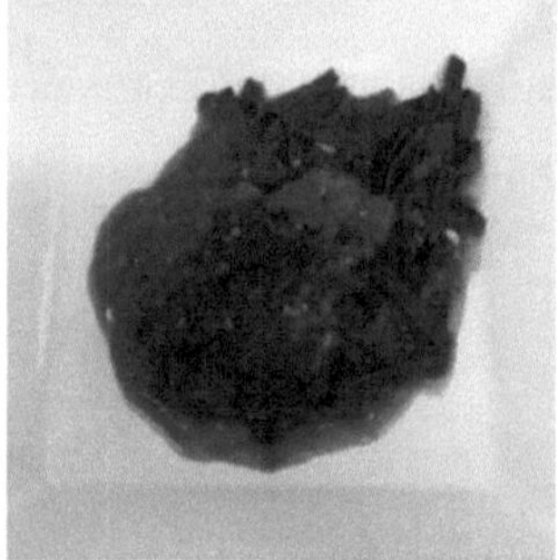

Figure 7 - Banana-green and ora-pro-nobis flour noodles with sauce (left) and without sauce (right) opened after vacuum-packing.

3.7. APPLICATION OF THE *SOUS VIDE* TECHNIQUE

The sample with sauce opened immediately after applying the technique showed positive organoleptic characteristics and cooking time, but the sample without sauce showed strands sticking together when opened at the same time as those with sauce (Figure 8).

Figure 8. Green banana flour and ora-pró-nobis noodles, with sauce (after applying the sous vide technique).

3.8. FINALISATION

The results for the samples with sauce kept frozen at <12°C for 5 days were satisfactory, showing pleasant visual and sensory characteristics; however, the samples without sauce showed little moisture, causing the strands to stick together (Figure 9). After 15 days of storage at <12°C, the results obtained were similar to those of the samples kept for only 5 days (Figure 10). However, at 30 days of storage at <12°C, the samples, both with sauce and without sauce, showed unsatisfactory results in terms of visual appearance compared to the other samples (Figure 11).

Figura 9.   Green banana and ora-pro-nobis flour noodles with sauce (left) and without sauce (right) after 5 days in the freezer.

Figura 10.   Green banana and ora-pro-nobis flour noodles with sauce (left) and without sauce (right) after 15 days in the freezer.

Figure 11. Green banana and ora-pro-nobis flour noodles with sauce (right) and without sauce (left) after 30 days under freezing.

CHAPTER 4

According to the Brazilian table of food composition (TACO) from the University of Campinas (Unicamp), traditional raw egg noodles have 10.3g of protein, 2.0g of lipids, 76.6g of carbohydrates and 371 kcal in 100g of pasta; however, the noodles developed with green banana flour and ora-pro-nóbis had 20.87% more protein than traditional pasta, corresponding to 16.6% of the total energy value of protein in a standard 2,000 kcal diet. The high protein content found in the pasta may be related to the addition of ora-pro-nóbis flour to its composition, which has a high content of the nutrient (ALMEIDA, 2014). According to the Resolution of the Collegiate Directorate - RDC No. 54 of 12 November 2012 of the National Health Surveillance Agency, a food considered to have a high protein value must contain at least 12g of protein in 100g of the product and quantities of essential amino acids that meet the established conditions.

However, the lipid content of the noodles added with green banana flour and ora-pro-nobis was high compared to traditional noodles, but free of trans fat, which can be justified based on the ingredients used to make them, such as olive oil and eggs, which are a source of lipids. Lipids are widely used in the food industry, as they provide aroma, flavour and palatability to foods (FoodIngredients Brasil, 2016), however, according to the updated Brazilian Guideline on Dyslipidemia and Prevention

According to the 2017 Atherosclerosis Study, it is recommended to use a diet free of trans fatty acids, with consumption of saturated fatty acids < 10 per cent of the total calorific value for healthy individuals and < 7 per cent of the total calorific value for those at increased cardiovascular risk.

In the study by Choo & Aziz 2010, the noodles formulated with green banana flour had a lower glycaemic index due to the amount of resistant starch present in the still green fruit according to the Von Loescke scale. The noodles made with green banana flour and ora-pro-nóbis had a lower carbohydrate content when compared to traditional noodles, 12.66% less. Despite this lower amount of carbohydrate, the product developed still has a high carbohydrate content, but the quality and complexity of the carbohydrate present differs from traditional noodles, given that green banana and ora-pro-nobis flour were added to the noodles and these have high amounts of fibre.

With a view to combining food quality with flavour, appearance, nutritional values and conservation, the stages of sous vide production were applied, starting with vacuum packaging (ARMSTRONG, 2003). Rocha et al. (2008) point out that products that are vacuum-packed using an appropriate sealing machine have a longer shelf life due to the total removal of oxygen from the packaging, thus minimising the growth of some aerobic microorganisms, such as Samonella and Listeria Monocytogenes.

According to Kolva (2003), the benefits of the sous vide technique for products are many, among them the quality of the consistency, making the food softer, preserving the nutritional value during cooking, and maintaining the original flavour. These aspects were observed in the green banana and ora-pro-nobis flour noodles which, when cooked directly in boiling water, were lighter in colour after cooking when compared to cooking in a sous vide thermocirculator which, due to the packaging, avoided the "detachment" of micronutrients from the food in the water.There is no consensus in the literature on the ideal cooking temperature for macaroni in a sous vide thermocirculator. Although this technique was developed to increase the preservation of fresh products, it is more commonly used with protein foods (RAMOS, 2004). For cooking the product, a temperature of 73°C was used in the thermocirculator, a value considered safe by the Primer on Good Practices for Food Services, Resolution - RDC No 216/2004, which states that all parts of the food should reach a minimum temperature of 70°C. Just as there is no ideal cooking temperature for macaroni in the sous vide thermocirculator, there is also no consensus in the literature regarding the product's shelf life, but it is suggested that these products can reach a shelf life of 18 months (RAMOS, 2004).

CHAPTER 5

FINAL CONSIDERATIONS

This work has provided a better understanding of Unconventional Food Plants and their rich heritage, which remains unknown and understudied. The use of one of these, as well as one of the world's most consumed fruits, the banana, but still in its green ripening phase, for the preparation of a common food product, such as pasta, brought together the opportunity to enrich it nutritionally with accessible foods known for their quality nutritional content. Pasta is a food consumed in a variety of ways all over the world, but it is rich in refined carbohydrates. In order to contribute to its nutritional improvement, the pasta was added with green banana flour and ora-pro-nóbis flour, known for being a good source of fibre (resistant starch) and protein, respectively. In the first stages of the product's development, satisfactory results were obtained in terms of colour, flavour, aroma and texture, achieving the initial objectives. The **sous vide** technique was applied in order to enhance the quality of the raw material and thus help maintain its nutrients as well as the flavour of the ora-pro-nóbis, which were preserved by vacuum cooking. After concluding this research, we can consider the product to be a good innovation that will combine technological advantages in order to preserve the nutrients present in the pasta. Finally, it is concluded that further tests need to be carried out in order to improve the **sous vide** technique **employed**, as well as characterising the fibre content, essential amino acids and micronutrients present, in addition to assessing its microbiological safety and analysing public acceptance in relation to organoleptic characteristics.

CHAPTER 6

BIBLIOGRAPHICAL REFERENCES

ABIMAP - **Brazilian Association of Biscuit, Pasta and Industrialised Bread & Cake Industries.** MACARRON: Learn more about this food. Available at: <http://www.abimapi.com.br/estatistica-massas.php>. Accessed on: 26 Feb. 2016.

ALMEIDA, M. E. F. et al.Chemical Characterisation of Non-Conventional Vegetables Known as Ora-Pro-Nobis. **Biosci. J.**, Uberlandia, v. 30, n. 1, p. 431-439, Jun. 2014. Available at: <http://www.seer.ufu.br/index.php/biosciencejournal/article/view/17555/14557>. Accessed on: 22 August 2017.

ARDISSONE, 2009. **BIODIVERSITY THROUGH THE MOUTH - Unconventional Food Plants (PANCs).** Extension Project: VIVEIRISM STUDIES AND PRACTICES IN A FARMERS' TRAINING CENTRE - Institute of Biosciences - UFRGS and Cooperfumos - MPA (Small Farmers' Movement) of the Bioenergy and Food Training Centre - São Francisco de Assis, Santa Cruz do Sul, RS. Available at: <http://www.ufrgs.br/viveiroscomunitarios/publicacoes/Cartilha%20Biodiversidade%20pela%20Boca.pdf> Accessed on: 02 Mar. 2016.

ARMSTRONG, G. A. **Sensory quality/ consumer acceptance of foods processed by the sous-vide system.** University of Ulster Online. Available at:http://www.science.ulst.ac.uk/sensory_Analysis.htm Accessed 23 Aug. 2017.

BALDWIN, D. Practical Guide to Sous-vide Cooking Written by Dr Douglas Baldwin. Gastronomia Lab. Available at: <http://www.douglasbaldwin.com/guia_pratico_sousvide.pdf>. Accessed on: 28 August 2017.

BATISTA, F. M.; RISSIN, A. **The** nutritional transition in Brazil: regional and temporal trends. **Cad. Saúde Pública**, n. 19, v. 1, p. 181-91, 2003. Available at: <https://www.nescon.medicina.ufmg.br/biblioteca/imagem/1531.pdf>. Accessed on 29 Feb. 2016.

BRAZIL. **Resolution of the Collegiate Directorate - RDC No. 54, of 12 November 2012.** Available at: <http://portal.anvisa.gov.br/documents/%2033880/2568070/rdc0054_12_11_2012.pdf/c5ac23fd-974e-4f2c-9fbc-48f7e0a31864>. Accessed on: 29 Aug. 2017.

Manual of non-conventional vegetables. Ministry of Agriculture, Livestock and Supply. Belo Horizonte, MG, 2010. p. 5-10. Available at: <http://www.abcsem.com.br/docs/manual_hortalicas_web.pdf> Accessed on: 30 Feb.

2016.

Primer on Good Practices for Food Services, Resolution -RDC n° 216/2004, page 33, Brasília, 3ª Edition. Available at: http://saude.es.gov.br/Media/sesa/NEVS/Alimentos/cartilha_gicra_final.pdf. Accessed on 25 August 2017.

EMBRAPA. BRAZILIAN AGRICULTURAL RESEARCH COMPANY. **Mixed Green Banana and Brazil Nut Flour.** Brasília, DF: Embrapa-SP, 2009. Available at: <http://ainfo.cnptia.embrapa.br/digital/bitstream/item/55500/1/AGROIND-FAM-Flour-mixed-banana-green-cast-for-ed01-2009.pdf>. Accessed on: 25 Feb. 2016.

ET-DASH, A.; CABRAL, L.C.; GERMANI, R. **Use of mixed wheat and soya flour in bread production.** Brasília: Brazilian Agricultural Research Corporation, 1994. 89 p. (Coleção Tecnologia de farinhas mistas, v.3). Available at: <http://livimagens.sct.embrapa.br/amostras/00060680.pdf>Accessed on: 02 March 2016.

Faludi AA, Izar MCO, Saraiva JFK, Chacra APM, Bianco HT, Afiune Neto A et al. **Update of the Brazilian Guideline on Dyslipidaemias and Prevention of Atherosclerosis -** 2017. Page 18. ArqBrasCardiol 2017; 109(2Supl.1):1-76. Available at: http://publicacoes.cardiol.br/2014/diretrizes/2017/02_DIRETRIZ_DE_DISLIPIDEM IA S.pdf. Accessed on 25 August 2017.

Green banana flour as a functional ingredient in food products. Ciência Rural, v.45, n.12, dec, 2015. Available at: http://www.beltnutrition.com.br/media/artigos/banana-verde-como-ingrediente- functional.pdf. Accessed on: 23 August 2017

FOOD AND AGRICULTURE ORGANISATION - FAO, EMBRAPA Empresa Brasileira de Pesquisa Agropecuaria, 2013. Available at: <https://www.embrapa.br/documents/1355135/0/Banana_Mundo_2013/a00fd130-94c9-43ba-a60f-0f6c1bbe0737>. Accessed 25 Feb. 2016.

FoodIngredients. **Lipids and their main functions** Brazil N° 37 -2016. Page 57. Available at: http://revista-fi.com.br/upload_files/201606/20160604926010014652395 02.pdf. Accessed on: 23 August 2017.

GARCIA, I. M. G.; HEBBEL, I. S. **Sous vide cooking method: effects on the beef shoulder cut.** Revista iniciação vol. 1 oct. n. 1, São Paulo, 2010. Available at: <http://www1.sp.senac.br/hotsites/blogs/revistainiciacao/wp-content/uploads/2013/07/10. pdf>. Accessed on 25 Feb. 2016.

ADOLFO LUTZ INSTITUTE. Physico-chemical methods for analysing food. São Paulo: Instituto Adolfo Lutz, 2008. p. 1020. Available at: <http://www.crq4.org.br/sms/files/file/analisedealimentosial_2008.pdf>. Accessed on: 29 August 2017.

BRAZILIAN INSTITUTE OF GEOGRAPHY AND STATISTICS (IBGE).
Household Budget Survey 2008-2009: **Anthropometry and nutritional status of children, adolescents and adults in Brazil.** Rio de Janeiro, v. 1, p. 11-130, 2010. Available at: <http://www.ibge.gov.br/home/estatistica/populacao/condicaodevida/pof/2008_2009_encaa/>. Accessed on: 11 Nov. 2016.

IBGE, EMBRAPA Empresa Brasileira de Pesquisa, 2013 Agropecuaria. Available at: <https://www.embrapa.br/documents/1355135/0/Banana_Brasil_2013/9027a417-ae4f-40d4-b088-cc79350568c>. Accessed 25 Feb. 2016.

KINUPP, V.F.; BARROS, I.B.I. Protein and mineral contents of native species, potential vegetables and fruits. **Ciência e Tecnologia de Alimentos.** v.28, n.4, p.846- 857, 2008. Available at: <http://www.scielo.br/scielo.php?script=sci_arttext&pid=S0101 - 20612008000400013&lng=pt&nrm=iso>. Accessed on 02 March 2016.

LICHTEMBERG,LICHTEMBERG. Banana Harvest and Post-Harvest. **Revista Informe Agropecuário.** V. 29, n. 245. Epamig, Belo Horizonte - MG, 2008. Available at: <http://wp.ufpel.edu.br/fruticultura/files/2011/10/pag029_036-Lecture214-11.pdf> Accessed on: 24 Feb. 2016.

LEWINSOHON, T. M.; PRADO, P. I. **Biodiversidade Brasileira: Síntese do Estado Atual do Conhecimento.** nov. 2000. Available at: <http://www.mma.gov.br/port/sbf/chm/doc/estarte.doc> Accessed on: 26 Feb. 2016.

MEDEIROS, D. L. **Research into the sous vide technique.** Monograph (specialisation) - University of Brasília, Centre of Excellence in Tourism, 2009. Available at: <http://bdm.unb.br/bitstream/10483/1042/1/2009_DalilaLorenyMedeiros.pdf>. Accessed on: 26 Feb. 2016.

PIRES, M. C. **Preparation and sensory analysis of gluten-free pastry with added green banana flour.** 2012. 33 f. Scientific Initiation Project (Undergraduate Degree in Pharmacy) - Sacred Heart University, Bauru, 2012

RAIMUNDO, M. G. M.; SÁ, A. C. E.; MICHELAZZO, L. A. **Nutritional composition and culinary use of bananas.** São Paulo: Biological Institute, 2013. Available at: <http://www.biologico.sp.gov.br/docs/livro_banana/capitulo14.pdf>. Accessed on 24 Feb. 2016.

RAMOS, A. E. A. **The sous vide system.** Monograph (specialisation) - University of Brasília, Centre of Excellence in Tourism, 2004. Available at: <http://bdm.unb.br/bitstream/10483/517/1/2004_AnaElisaAguiarRamos.pdf>. Accessed on: 25 Feb. 2016.

RAMOS, P.D.; LEONEL, M.; LEONEL, S. Resistant Starch in Green Banana Flour. **Rev.Alim. Nutr.**, Araraquara, v.20, n.3, p. 479-483, 2009. Available at: <http://serv-

bib.fcfar.unesp.br/seer/index.php/alimentos/article/viewFile/1151/846>. Accessed on: 25 Feb. 2016

Resolution RDC no. 93, of 31 October 2000. Ementa: **Provides for the Technical Regulation for Establishing the Identity and Quality of Pasta**: <http://portal.anvisa.gov.br/wps/wcm/connect/59cd1a004745896b9384d73fbc4c6735/RDC_93_2000.pdf?MOD=AJPERES> Accessed on: 29 Feb. 2016.

ROCHA et al. **Noodles with** dehydrated **ora-pro-nobis** (PereskiaaculeataMiller). Alimentos e Nutrição, v.19, n.4, p.459-465, 2008. Available at: <http://servbib.fcfar.unesp.br/seer/index.php/alimentos/article/viewFile/656/ 552>. Accessed on: 25 Feb. 2016.

BRAZILIAN SUPPORT SERVICE FOR MICRO AND SMALL ENTERPRISES. **Banana**: SEBRAE/ ESPM **market studies**. [S.l.]: SEBRAE, 2008 (Market series). Available at: <http://201.2.114.147/bds/bds.nsf/8E2336FF6093AD96832574DC0045023C/$File/NT0003904A.pdf>. Accessed on: 24 Feb. 2016.

SILVA, M.C.A. **Technical and economic analysis of apple banana (Musa spp.) cultivation in the north-western region of the state of São Paulo.** 2004. 73 f. Dissertation (master's degree) - Universidade Estadual Paulista, Faculdade de Engenharia de Ilha Solteira, 2004. Available at: <http://hdl.handle.net/11449/98842. Accessed on: 28 August 2017.

SOUZA, J. M. L. et al. **Mixed Green Banana and Brazil Nut Flour.** Brasília-DF: Embrapa Informação Tecnológica, 2009. 49p. Available at: https://www.embrapa.br/busca-de-publicacoes/-/publicacao/975061/farinha-mista-de-banana-verde-e-de-castanha-do-brasil. Accessed on: 28 August 2017.

TACO: Brazilian Food Composition Table / NEPA - Food Studies and Research Centre. UNICAMP. - 4. rev. and ampl. ed. -- Campinas: 2011. 161 p. Available at: http://www.unicamp.br/nepa/taco/contar/taco_4_. Accessed on: 29 Feb 2016.

APPENDIX A - Weighing green bananas

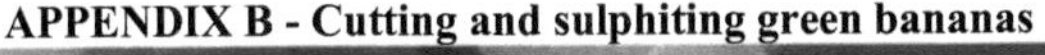

APPENDIX B - Cutting and sulphiting green bananas

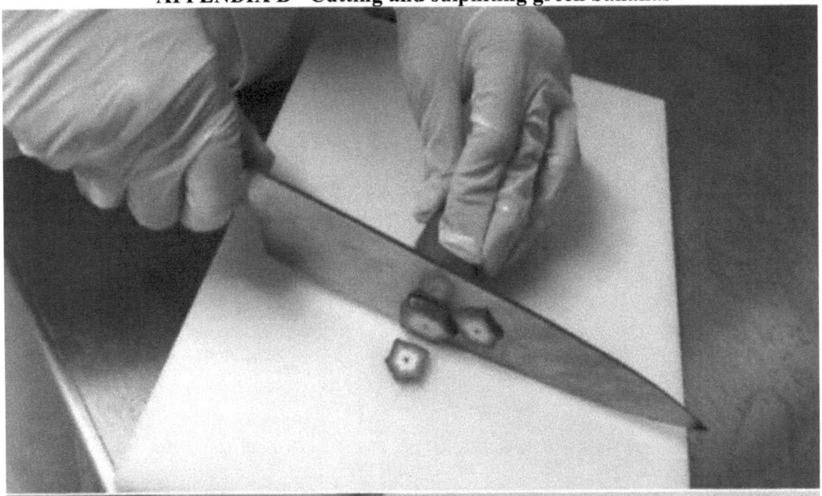

APPENDIX E - Leaves on screens suitable for kiln drying

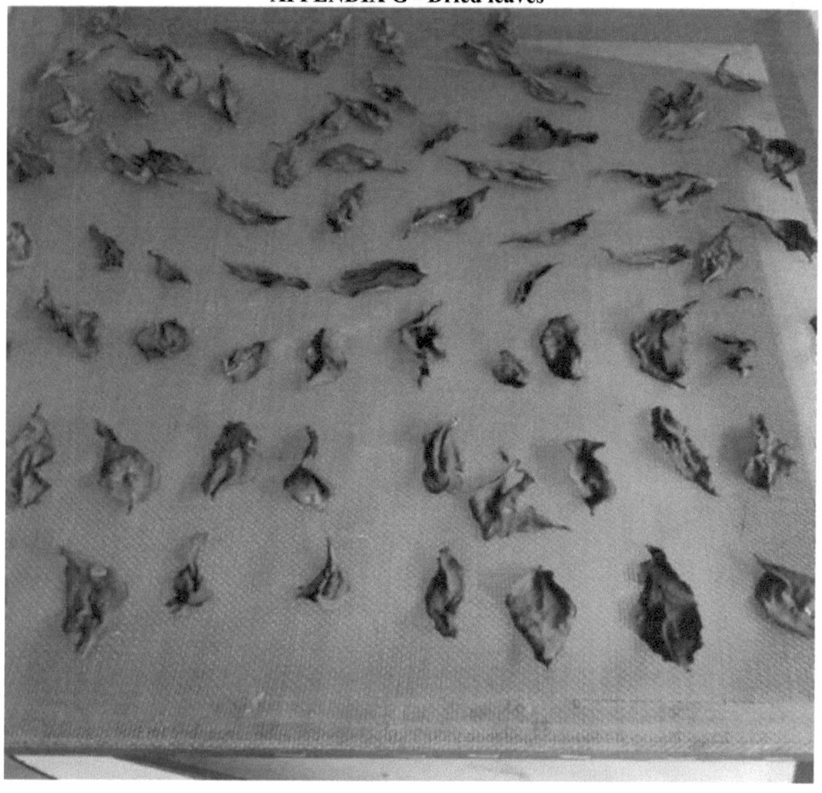